Managing Developers in the Age of Entitlement: A Tech Leader's Guide

Contents

Part 1: Understanding the Landscape

In the world of technology and project management, understanding the landscape in which we operate is as crucial as the skills and tools we employ. The tech industry, known for its rapid innovation and dynamic changes, has seen a significant shift in the attitudes and behaviors of its workforce, particularly among developers. This shift, often labeled as 'entitlement', has become a topic of keen interest and concern for project managers and team leaders who strive to steer their teams towards success while navigating the complexities of modern work environments.

The inception of this change can be traced back to the practices and cultures established by the FAANG companies (FAANG is an acronym for five of the best-performing tech-centric stocks of the past decade: Facebook, Amazon, Apple, Netflix and Google.) These industry giants, celebrated for their breakthroughs and market dominance, have inadvertently set a precedent in workplace expectations and employee attitudes. Their strategies in recruitment, compensation, and work environment have created a ripple effect across the tech sector, influencing not just the startups aspiring to emulate their success but also the established companies trying to compete for talent in a talent-driven market. In recruiting circles, the best and the brightest developers are often known as Unicorns because of the money that can attract. The high salaries, luxurious perks, and unique office cultures that define these companies have contributed to a sense of exceptionalism among their employees, particularly the developers, who often find themselves at the heart of these organizations' successes.

This exceptionalism has its pitfalls. The culture of entitlement, stemming from being part of a select group that drives innovation, has raised several challenges. It has affected how developers view their roles, their expectations from their employers, and their interactions within their teams. The belief in one's indispensability or superiority can disrupt team dynamics, hinder collaboration, and create an environment where collective goals may be overshadowed by individual egos. This is particularly challenging in a field like software development, where teamwork, communication, and a shared vision are paramount for the success of complex projects.

The influence of the FAANG companies extends beyond their direct employees. It sets a standard that affects the entire industry's talent pool. New graduates and aspiring developers enter the job market with expectations molded by the stories and myths surrounding these tech behemoths. This expectation mismatch can lead to job dissatisfaction and a constant quest for better opportunities, further fueling the cycle of entitlement and instability in the workforce.

To navigate this landscape, it is essential to understand not just the overt manifestations of entitlement but also its subtle undercurrents. The sense of being special or above average, the need for constant validation and recognition, and the resistance to mundane but necessary aspects of the job like documentation and adhering to processes, are all indicators of this mindset. These traits, while benign in isolation, can collectively hinder the growth and efficiency of teams.

In the subsequent chapters of this part, we delve deeper into the origins, manifestations, and impacts of this entitlement culture. From the roots of entitlement in the tech industry to the psychological underpinnings of this mindset, we explore how the culture fostered by the FAANG tech companies has permeated different aspects of software development and project management. We will also examine methodologies like Scrum, which, while designed to enhance productivity and collaboration, can sometimes exacerbate the problem by giving too much autonomy without adequate accountability. The focus then shifts to the phenomenon of the 'Snowflake Syndrome', a byproduct of extreme individualism, where the need to be seen as unique and exceptional undermines team cohesion and collective goals.

This exploration is not just an academic exercise but a necessary step for project managers and team leaders who are grappling with these challenges daily. By understanding the landscape, we lay the groundwork for developing effective strategies and tools to address these issues in the following sections of the book. It's about redefining the ethos of tech teams, realigning priorities, and fostering an environment where talent is nurtured, but not at the cost of teamwork and shared objectives.

The Golden Cage: How the FAANG Tech Giants Cultivated Developer Entitlement

In the realm of technology and software development, the term 'entitlement' has emerged as a buzzword, often thrown around in discussions about workplace culture, especially concerning developers. This entitlement, a complex blend of expectations, attitudes, and behaviors, has been significantly shaped by the practices and policies of the Big 5 tech giants. These companies, known for their groundbreaking innovations and dominant market presence, have inadvertently crafted an environment where developer entitlement has not just taken root but flourished.

At the core of this entitlement culture lies the unique workplace environment fostered by these tech behemoths. Renowned for their lavish campuses, stocked with amenities ranging from gourmet cafeterias to on-site health services, these companies have set a new standard for what a workplace can offer. This luxurious environment, while aimed at attracting top talent and boosting productivity, has also contributed to a growing expectation among developers for above-average workplace benefits, often leading to a sense of entitlement.

The recruitment strategies of these companies further amplify this phenomenon. In a fiercely competitive market for top tech talent, these giants engage in aggressive headhunting, offering substantial salaries and stock options. This practice, while beneficial in securing the best minds, has escalated salary expectations industry-wide, leading to a wage inflation that smaller companies often struggle to match. Developers, especially those in the early stages of their careers, begin to equate their self-worth with the size of their paychecks, a mindset that can lead to entitlement.

The prestige associated with working for one of the Big 5 has its own implications. Developers employed by these companies often enjoy a status akin to that of celebrities in the tech world. This status, coupled with the high-stakes projects they work on, fosters a sense of exceptionalism. Developers begin to see themselves as indispensable, a belief that can sometimes lead to a dismissal of criticism, resistance to feedback, and an overall sense of superiority over peers in other organizations.

This entitlement culture is not without its consequences. It risks creating a workforce that is less adaptable, less collaborative, and more focused on individual gains than collective achievements. For project managers and team leaders, this presents a significant challenge. Balancing the need to maintain an attractive work environment with the necessity of fostering a culture of humility and teamwork becomes a tightrope walk. The risk is a workforce divided, with a gap between those who demand exceptional treatment and those who focus on the collective good of the team and the project.

This divide extends beyond individual companies, permeating the entire tech industry. Startups and mid-sized companies find themselves in a tough spot, unable to match the salaries and perks offered by the Big 5, yet needing talented developers to drive their innovations. This disparity can lead to a talent drain, where the best minds are consistently lured away by the promise of higher salaries and better benefits, leaving a void that is hard to fill.

The long-term effects of this entitlement culture are far-reaching. It threatens to stifle innovation in smaller companies, create wage disparities that are unsustainable, and cultivate a workforce that is increasingly difficult to satisfy. The challenge for the tech industry is to find a balance – to provide an environment that attracts and retains talent, rewards innovation and hard work, but also keeps in check the entitlement that can hinder collaboration, stifle growth, and create a divisive workplace culture.

The entitlement culture fostered by the Big 5 tech giants is a multifaceted issue with deep-rooted implications for the tech industry. It requires a nuanced approach, one that understands the allure of such an environment, but also recognizes the need for a culture that values teamwork, adaptability, and a shared vision for success.

For the tech industry to continue thriving, it must navigate this landscape carefully, crafting policies and practices that foster a healthy, collaborative, and innovative work environment.

Scrum's Influence: Agile or Fragile?

The Scrum methodology, renowned for its agility and flexibility, has become a cornerstone in the world of software development. Its influence on developer behavior, team dynamics, and the overall work environment is profound. However, with its increasing adoption, a critical question arises: Does Scrum foster a stronger, more cohesive team ethos, or does it inadvertently contribute to a sense of entitlement among developers? This chapter delves into this paradox, examining the nuances of Scrum's impact on the modern software development team.

Scrum, at its core, is designed to empower teams, promoting autonomy, accountability, and a collaborative spirit. Its iterative nature and emphasis on self-organization aim to drive efficiency and innovation.

However, these very strengths of Scrum can sometimes morph into vulnerabilities, particularly when it comes to developer entitlement. The autonomy afforded by Scrum, while intended to boost initiative and creativity, can sometimes lead to a silo mentality where individual perspectives and priorities overshadow the collective goals of the team. This autonomy, in the absence of adequate checks and balances, can breed an environment where developers feel entitled to pursue their vision, often at the expense of collaborative synergy.

The dynamics of Scrum teams also play a critical role in shaping this environment. Scrum promotes flat hierarchies and self-organization, principles that are fundamentally empowering. However, in practice, this can lead to challenges in leadership and authority. Developers, emboldened by the lack of a traditional hierarchical structure, might feel more inclined to challenge decisions or resist guidance, perceiving it as an infringement on their autonomy. This resistance is not necessarily born out of defiance but rather out of a heightened sense of ownership and entitlement to one's ideas and methodologies.

Another aspect where Scrum's influence is markedly felt is in its iterative feedback loops. Continuous feedback, a pillar of Scrum, is intended to foster a culture of constant improvement and adaptability. However, the frequency and nature of this feedback can sometimes have unintended consequences. For developers, the regular scrutiny and need to adapt rapidly to feedback can foster a defensive mindset, one where criticism, however constructive, is viewed as a personal affront. This defensiveness can be a breeding ground for entitlement, as developers might begin to view their approaches and solutions as beyond reproach, aligning feedback with personal criticism rather than opportunities for growth.

Furthermore, Scrum's emphasis on sprint goals and deliverables can sometimes overshadow the importance of long-term vision and strategic planning. Developers might become so entrenched in the immediacy of sprint objectives that they lose sight of the broader project goals or organizational objectives. This myopic focus can lead to a sense of entitlement regarding one's contributions, with developers feeling disproportionately important to the project's immediate successes, often neglecting the collaborative effort and the larger picture.

The cumulative effect of these factors is a complex environment where entitlement can quietly take root. This entitlement is not always overt; it can manifest in subtle ways, such as resistance to certain tasks, insistence on particular methodologies, or reluctance to fully embrace team decisions. It can also lead to a fragmented team dynamic, where the collective spirit of Scrum is overshadowed by individual agendas and self-interest.

In addressing these challenges, the key lies in striking a balance. It involves reinforcing the principles of Scrum - collaboration, flexibility, and continuous improvement - while also instituting mechanisms that keep entitlement in check. This balance requires a nuanced understanding of Scrum's philosophy, an appreciation of the individual strengths of developers, and a commitment to fostering a team culture that values collective success over individual accolades.

While Scrum offers a robust framework for agile development, its impact on developer behavior and entitlement is multifaceted. Navigating this landscape requires a thoughtful approach, one that leverages Scrum's strengths to build a cohesive, dynamic, and balanced team environment. By doing so, project managers and team leaders can harness the full potential of Scrum, mitigating the risks of entitlement and fostering a culture of shared success and continuous growth.

Snowflake Syndrome in Software Development

In software development, the emergence of what is colloquially termed as the 'Snowflake Syndrome' presents a unique set of challenges. This phenomenon, characterized by over-individualism among team members, has distinct origins and significant impacts on teamwork and project outcomes. The term 'Snowflake Syndrome' in this context refers to a tendency among certain developers to perceive themselves as unique and indispensable, often leading to a reluctance to conform to standardized processes or collaborate effectively with others. This chapter explores the definition, origins, and implications of this syndrome in the context of software development teams.

The origins of the Snowflake Syndrome in tech teams can be traced back to a combination of factors. The tech industry's rapid growth and evolution have created an environment where exceptional talent is highly prized. This environment often fosters a culture that celebrates individual achievement and innovation. While this can drive progress and creativity, it can also lead to a sense of over-individualism, where developers may start to feel that their skills and contributions are so unique that standard rules and processes do not apply to them. This mindset is further reinforced by the industry's competitive nature, where being seen as distinctive can be perceived as a route to career advancement and recognition.

Educational backgrounds and training also play a significant role in nurturing this mindset. Many developers come from academic environments that emphasize personal achievement and technical prowess. While this focus on individual skill development is critical, it can sometimes overshadow the importance of teamwork and collaboration, skills that are equally essential in the professional world. This gap in collaborative training often manifests in the workplace, where developers struggle to transition from an individual-centric approach to a team-oriented one.

The impact of the Snowflake Syndrome on software development teams and project outcomes can be profound. One of the most immediate effects is on team dynamics. Teams with members who exhibit strong over-individualism may experience difficulties in communication and collaboration. These difficulties arise because team members may prioritize their ideas and approaches over others, leading to conflicts, a lack of cohesion, and an overall inefficient team dynamic. This situation can be particularly challenging in agile environments, where collaboration and adaptability are key to success.

Project outcomes are also significantly affected by this syndrome. When team members are unwilling to align with standardized processes or collaborate effectively, it can lead to inconsistencies in the quality of the work, delays in project timelines, and even failure to meet project objectives. In software development, where projects often require a high degree of coordination and integration of different components, the inability to work cohesively can have detrimental effects on the final product.

Addressing the Snowflake Syndrome requires a multifaceted approach. One of the key strategies is to foster a culture that values and rewards teamwork and collaboration as much as individual achievement. This cultural shift can be facilitated through team-building activities, collaborative projects, and recognition of team successes. Leadership and management also play a critical role in mitigating this syndrome. Effective leaders need to set clear expectations around teamwork, provide opportunities for collaborative skill development, and ensure that processes and standards are adhered to without stifling individual creativity.

Another important aspect is to provide training and development opportunities that focus on soft skills, such as communication, empathy, and collaboration. These skills are essential for creating an environment where team members can work together effectively, respecting each other's contributions while working towards a common goal.

While the Snowflake Syndrome presents significant challenges in software development teams, it can be addressed through a combination of cultural shifts, effective leadership, and focused training. By tackling the issue of over-individualism, teams can enhance their collaboration, improve their dynamics, and ultimately achieve better project outcomes. Recognizing the value of each team member's unique contributions while emphasizing the importance of collective effort is key to overcoming the challenges posed by this syndrome.

Part 2: Leadership and Management Strategies

In Part 2 we delve into the essential role of leadership and management in the realm of software development, especially in the context of the evolving challenges discussed in Part 1. This section is pivotal, marking a transition from understanding the landscape of developer entitlement and individualism to actively addressing these challenges through effective leadership and management techniques. Here, we explore the intricate dynamics of leading and managing tech teams, focusing on how to cultivate a harmonious, productive, and collaborative work environment amidst the complexities of modern tech culture.

Leadership in software development transcends beyond mere technical expertise or project management. It encompasses the art of navigating complex human dynamics, understanding the diverse motivations and personalities within a team, and aligning individual ambitions with collective goals. This part of the book is dedicated to unraveling various leadership strategies and management approaches that are crucial in counteracting the growing sense of entitlement and individualism among developers. It provides a roadmap for leaders to exert their influence effectively, shaping teams that are not only technically proficient but also resilient, adaptable, and united.

At the core of this discussion is the art of communication. Effective leadership is fundamentally rooted in how leaders communicate with their teams. This involves not just imparting instructions or ideas but also engaging in active listening, displaying empathy, and understanding and addressing team members' needs and concerns. In this section, readers will gain insights into refining their communication skills, which are pivotal in resolving conflicts, building trust, and fostering an environment of open, productive dialogue.

Another critical theme explored here is striking the right balance between granting autonomy and ensuring accountability. In agile and dynamic tech environments, autonomy is often a key driver of innovation and job satisfaction. However, without proper guidance and accountability, autonomy can give way to the challenges of entitlement and detachment from team objectives. This section discusses strategies to empower developers effectively while maintaining a cohesive and goal-oriented team structure.

Leadership and management in tech are about creating a culture where individual talents are recognized and nurtured, but not at the expense of the team's collective success. It's about building a framework where each member feels valued and motivated to contribute to the larger vision of the project and the organization. In these chapters, we explore how to achieve this delicate balance, providing leaders and managers with the tools and insights needed to foster a healthy, productive, and engaging work environment.

Effective Communication with Entitled Developers

Effective communication with entitled developers requires a nuanced approach that combines understanding, empathy, and assertiveness. In the world of software development, where teamwork and collaboration are essential, addressing the unique challenges posed by entitlement can significantly impact the overall productivity and harmony of the team. This chapter explores strategies for effectively communicating with developers who may display a sense of entitlement, focusing on understanding their perspectives and motivations and guiding them towards a more collaborative and productive mindset.

The first step in effective communication with entitled developers is to understand the root of their entitlement. Entitlement often stems from a variety of sources, including the highly competitive nature of the tech industry, the elevated status of developers in the modern workplace, and sometimes, personal traits or past experiences. By understanding these underlying factors, a manager or team leader can tailor their approach to address the specific needs and concerns of the individual, rather than applying a one-size-fits-all solution.

One effective strategy is to establish a rapport based on respect and mutual understanding. This involves acknowledging the developer's skills and contributions while also setting clear expectations regarding teamwork and collaborative practices. It's important to communicate that while individual talents are valued, the success of the project and the team as a whole takes precedence. This approach can help in reducing resistance and opening up a dialogue based on mutual respect and common goals.

Active listening plays a crucial role in this process. It involves giving developers the space to express their views and concerns without immediate judgment or dismissal. This not only helps in understanding their perspective but also demonstrates a willingness to engage in a constructive dialogue. Active listening can often reveal the deeper issues behind entitled behavior, such as a need for recognition, fear of being undervalued, or concerns about career progression.

Feedback is another critical component of effective communication with entitled developers. Feedback should be regular, constructive, and focused on behaviors rather than personal traits. It's important to frame feedback in a way that emphasizes growth and development, rather than criticism. For example, instead of pointing out a failure to collaborate, suggest ways in which collaboration could enhance their individual contribution and the project's success.

It is also essential to lead by example. Demonstrating the values of humility, teamwork, and open communication sets a tone for the entire team. When leaders exhibit these traits, it encourages developers to follow suit, creating a more cohesive and collaborative environment.

Additionally, setting clear goals and expectations is vital. This involves not only defining what needs to be achieved but also explaining why it is important. When developers understand the rationale behind certain processes and decisions, they are more likely to buy into them and participate actively.

Conflict resolution skills are also crucial when dealing with entitled behavior. Conflicts, if not managed properly, can escalate and affect team morale and productivity. It's important to address conflicts promptly and fairly, ensuring that all parties feel heard and that solutions are focused on the best interest of the team and project.

Effectively communicating with entitled developers is about striking a balance between recognizing and valuing individual talents while also emphasizing the importance of collaboration and team success. It requires understanding, empathy, active listening, constructive feedback, leading by example, setting clear goals and expectations, and proficient conflict resolution. By employing these strategies, leaders and managers can guide entitled developers towards a more collaborative, productive, and positive work environment.

Leadership Techniques for Tech Environments

Leadership in technology environments demands a unique blend of flexibility, understanding, and assertiveness. Given the diverse range of personalities and the high-value placed on creativity and innovation in tech teams, leaders must adapt their styles to effectively manage and guide their teams. The key lies in balancing the need to foster creativity and innovation with the necessity to maintain discipline and focus on the collective goals of the team and the project.

Understanding the individual personalities within a tech team is the first step towards effective leadership. Developers come from various backgrounds and possess different sets of skills, motivations, and work ethics. Some may thrive in highly structured environments, while others might find their creativity sparked in more flexible settings. A successful leader recognizes these differences and adapts their approach accordingly. This might involve providing more structured guidance to some team members while granting others the autonomy they need to be creative. The goal is to create an environment where all team members feel supported and motivated to contribute their best work.

Communication is a critical tool in achieving this balance. Leaders must be clear in their expectations and transparent in their decision-making processes. Regular team meetings, one-on-one check-ins, and open lines of communication help in ensuring that all team members are aligned with the team's goals and are fully aware of their roles and responsibilities. This level of clarity helps in minimizing misunderstandings and conflicts, thereby maintaining a disciplined yet creative work environment.

Another aspect of effective leadership in tech environments is the ability to inspire and motivate. This often involves more than just providing technical guidance. Leaders need to connect with their team members on a personal level, understanding their career aspirations, and helping them see how their work contributes to the larger goals of the project and the organization. This connection can be a powerful motivator, driving developers to not only meet but exceed expectations.

However, maintaining discipline is equally important. This does not necessarily mean imposing strict rules or micromanaging, but rather setting clear standards for quality, performance, and collaboration. It involves holding team members accountable for their work and ensuring that everyone contributes equally to the team's efforts. Discipline in this context is about creating a culture of responsibility and professionalism where deadlines are met, quality standards are upheld, and team members are reliable and respectful of each other's contributions.

Adapting to the ever-changing landscape of technology is another critical component of leadership in tech environments. This means staying abreast of new technologies, methodologies, and industry trends, and being open to changing processes and workflows as needed. Leaders who are flexible and open to change can better guide their teams through transitions, ensuring that they remain competitive and innovative.

Conflict resolution is an inevitable part of managing diverse teams, and leaders must be adept at navigating and resolving conflicts in a way that is fair and constructive. This involves listening to all sides, understanding the root causes of the conflict, and working collaboratively to find solutions that are in the best interest of the team and the project.

Leading tech teams requires a multifaceted approach. It demands an understanding of the diverse personalities within the team, effective communication, the ability to inspire and motivate, maintaining discipline, staying adaptable to change, and proficient conflict resolution skills. By mastering these techniques, leaders can create an environment where creativity flourishes, discipline is maintained, and the team works cohesively towards achieving their collective goals.

Hard Choices: When Firing Becomes a Necessary Solution

The decision to terminate an employee is one of the most challenging and consequential actions a leader can take.

It's a decision that not only affects the individual involved but also has significant implications for the team and the organization as a whole. In the context of software development, where teams often work in closely-knit units, the impact of firing an employee can be particularly profound. This chapter delves into the complexities surrounding the termination of employees, focusing on the legal and ethical considerations, as well as the impact on team morale.

The legal considerations of firing an employee are paramount. It's essential to ensure that the termination process is compliant with employment laws and regulations. This compliance involves giving fair warning, documenting performance issues, and providing a clear rationale for the decision. Failure to adhere to these legal requirements can result in litigation, financial penalties, and damage to the company's reputation. Therefore, it's crucial for leaders to be well-versed in the legal aspects of employment and termination, often necessitating consultation with human resources or legal professionals.

Ethical considerations are equally important. Firing an employee should always be a last resort, after all other avenues such as coaching, mentoring, and performance improvement plans have been exhausted. The process should be handled with dignity and respect, ensuring that the employee is given a fair and honest assessment of their performance and behavior. Leaders must balance the needs of the organization with compassion and empathy for the individual involved. The way a termination is handled can significantly impact the remaining team members' perception of the organization and its leadership.

Maintaining team morale in the aftermath of a termination is another critical aspect. The departure of a team member, especially under such circumstances, can create uncertainty and anxiety among the remaining employees. They may question the stability of their own positions or the overall health of the organization. To address these concerns, it's essential for leaders to communicate openly and transparently with the team. This communication should include the reasons for the termination (while respecting the privacy of the individual involved) and how the decision aligns with the broader goals and values of the organization.

Leaders should also be proactive in managing the team's dynamics post-termination. This management involves redistributing responsibilities, possibly bringing in new talent, and ensuring that the team has the resources and support needed to move forward. It's an opportunity to reaffirm the team's objectives and to reinforce a culture of performance, collaboration, and mutual respect.

Moreover, a termination can serve as a learning experience for both the leadership and the team. It highlights the importance of clear expectations, regular feedback, and proactive management of performance issues. It's an opportunity to reflect on the organization's hiring practices, onboarding processes, and the effectiveness of its performance management systems.

In conclusion, the decision to fire an employee is never easy and comes with a host of legal, ethical, and morale considerations. It requires a careful, thoughtful approach, balancing the legal requirements with a compassionate handling of the individual involved. Effective communication, both during and after the process, is crucial in maintaining team morale and trust in leadership. By handling terminations with fairness, respect, and transparency, leaders can navigate these difficult situations in a way that upholds the values of the organization and supports the long-term health of the team.

Beyond Code: Getting Developers Engaged with Administrative Tasks

In the fast-paced world of software development, engaging developers in non-development tasks such as documentation and time tracking is often a challenge. These tasks, while crucial for the smooth running and success of projects, are frequently viewed as tedious or secondary by developers who are primarily focused on the creative aspects of coding. However, the importance of these administrative tasks cannot be understated, as they ensure project clarity, continuity, and accountability. This chapter explores methods to encourage developers to participate actively in these essential but often overlooked aspects of their work.

The first step in engaging developers with administrative tasks is to emphasize their importance and relevance to the overall success of projects. This communication involves clearly outlining how these tasks contribute to better project management, client satisfaction, and the quality of the final product. When developers understand the value and impact of their participation in documentation and time tracking, they are more likely to engage with these tasks willingly and diligently.

One effective method to encourage participation is to integrate these administrative tasks into the natural workflow of the development process. For instance, embedding documentation as a part of the coding process or incorporating time tracking within the tools developers already use can make these tasks feel less intrusive and more a part of their regular activities. Automation tools can also be employed to simplify these tasks, minimizing the time and effort required to complete them.

Another strategy is to tailor the approach to documentation and time tracking to suit the preferences and strengths of individual developers. Some may prefer detailed, comprehensive documentation, while others might be more inclined towards concise, bullet-point summaries. Similarly, different time tracking methods can be employed, from automated time tracking software to more manual, but straightforward approaches. Offering flexibility and options allows developers to choose the methods that work best for them, increasing their willingness to engage in these tasks.

Recognition and rewards can also play a significant role in motivating developers. Acknowledging the effort put into well-maintained documentation or accurate time tracking can reinforce the value of these tasks. This recognition can be in the form of verbal praise, public acknowledgment in team meetings, or even tangible rewards for teams or individuals who consistently excel in these areas.

Training and education are also crucial. Providing training sessions or resources on effective documentation practices and efficient time tracking can equip developers with the skills and knowledge they need to perform these tasks effectively. This training not only improves the quality of the output but also helps in reducing the perceived burden of these tasks.

Moreover, leading by example is one of the most powerful tools in a leader's arsenal. When team leaders and project managers actively engage in and prioritize documentation and time tracking, it sets a precedent for the rest of the team. This leadership shows that these tasks are not just for developers but are a shared responsibility and an integral part of the team's culture.

Finally, feedback loops are essential. Regularly seeking feedback from developers on how the processes of documentation and time tracking can be improved shows that their input is valued and that the organization is committed to making these tasks as efficient and painless as possible. This feedback can lead to continuous improvement in the way these tasks are approached and managed.

The Manager's Growth: Enhancing Emotional Intelligence and Resilience

The growth of a manager, particularly in the demanding and fast-paced environment of technology, hinges significantly on personal development. Key aspects of this development include emotional intelligence, resilience, and effective communication. These elements are not just ancillary skills; they are critical to navigating the complex landscape of team management and project leadership in tech.

Emotional intelligence (EQ) is the cornerstone of effective leadership. It involves the ability to understand, use, and manage one's own emotions in positive ways to relieve stress, communicate effectively, empathize with others, overcome challenges, and defuse conflict. In the context of tech management, this translates to a deeper understanding of team dynamics and individual motivations. A manager with high EQ can read the undercurrents of team interactions and individual behaviors, allowing for more effective management of diverse personalities and work styles.

Developing emotional intelligence starts with self-awareness. This means being cognizant of one's emotional states and how they influence thoughts and actions. Self-regulation follows, entailing the control of impulsive feelings and behaviors, managing emotions in healthy ways, taking initiative, following through on commitments, and adapting to changing circumstances. Empathy, an essential component of EQ, allows managers to understand the emotional makeup of their team members and treat them according to their emotional reactions. This empathy significantly impacts conflict resolution, motivation, and fostering a positive team environment.

Resilience in management is the ability to bounce back from setbacks, adapt well to change, and keep going in the face of adversity. In the tech world, where the only constant is change, resilience is a critical trait for managers. It involves maintaining a level of flexibility and keeping a positive attitude, even when things don't go as planned. Resilient managers are able to view challenges as opportunities to learn and grow rather than insurmountable obstacles.

Building resilience involves maintaining a positive outlook, seeing the big picture, and focusing on what can be controlled. It also includes taking care of one's physical and emotional health, as a healthy body can often mean a healthy mind. Moreover, resilient managers are adept at building and maintaining strong support networks, both professionally and personally. These networks provide a sounding board and a source of comfort and guidance during challenging times.

Effective communication is another pillar of personal development for managers. It's about more than just transmitting information; it's about understanding the emotion and intentions behind the information. Effective communication can defuse a situation, create mutual understanding and respect, and lead to a shared solution to problems. For tech managers, this means being able to clearly articulate project goals, deliver constructive feedback, and communicate complex technical information to non-technical stakeholders.

To enhance communication skills, managers should practice active listening, which involves fully concentrating, understanding, responding, and then remembering what is being said. They should also be clear and concise in their communication, avoiding unnecessary jargon or ambiguity. Additionally, being open to feedback and willing to engage in two-way communication can significantly improve a manager's communication effectiveness.

Balanced Perspectives: A Balanced View on Developer Entitlement

In the landscape of modern software development, the concept of developer entitlement often carries a predominantly negative connotation, conjuring images of demanding, inflexible individuals who place their needs and opinions above those of the team. However, a nuanced understanding of this phenomenon reveals that entitlement, like many other traits, has both positive and negative aspects. To manage teams effectively, especially in the dynamic and fast-paced tech industry, it is crucial to adopt a balanced perspective on developer entitlement, recognizing its potential benefits and drawbacks.

On the positive side, entitlement in developers can stem from a strong sense of self-worth and confidence in their skills and contributions. This self-assurance, when properly channeled, can lead to significant innovations and breakthroughs. Developers who believe strongly in their capabilities are often willing to push boundaries, explore new technologies, and challenge the status quo.

This can be a powerful driver of progress in an industry that thrives on innovation.

Furthermore, entitlement can sometimes be a manifestation of a developer's deep commitment to quality and excellence. Developers who take great pride in their work may exhibit signs of entitlement, not out of arrogance, but from a genuine desire to produce the best possible outcomes. They may resist shortcuts or subpar solutions, insisting instead on standards and approaches that align with their high personal and professional benchmarks.

However, the negative aspects of developer entitlement cannot be ignored. It can lead to conflicts within teams, create barriers to collaboration, and ultimately hinder project progress. Entitlement can manifest in an unwillingness to compromise or consider others' opinions, a resistance to feedback, and an expectation of special treatment. This can disrupt team dynamics, create an unhealthy work environment, and even impact the overall success of projects.

To manage developer entitlement effectively, it is crucial to strike a balance between leveraging its positive aspects and mitigating its negative impacts. This involves acknowledging and valuing the strengths and contributions of individual developers while fostering a culture of teamwork and collaboration. Leaders and managers must set clear expectations around team behavior and project goals, ensuring that individual aspirations align with collective objectives.

Communication plays a pivotal role in managing entitlement. Open, honest, and respectful dialogue can help in addressing issues related to entitlement. It is important for leaders to provide constructive feedback, recognizing and reinforcing positive behaviors while addressing negative ones. This feedback should be specific, objective, and focused on behaviors rather than personal attributes.

Creating opportunities for professional growth and development can also help in mitigating negative entitlement. By providing challenges, learning opportunities, and pathways for career advancement, organizations can channel developers' ambitions and sense of self-worth into productive and positive outcomes. This not only aids in personal and professional development but also benefits the organization by fostering a more engaged and motivated workforce.

In conclusion, a balanced view of developer entitlement is essential for effective team management in the tech industry. Recognizing that entitlement can be both a strength and a liability allows leaders to harness its positive aspects while mitigating its negative impacts. Through clear communication, constructive feedback, and opportunities for growth, it is possible to create an environment where individual talents are celebrated, but not at the expense of teamwork, collaboration, and the collective success of the team.

Part 3: Fostering Growth and Collaboration

Part 3 of this book, "Fostering Growth and Collaboration," shifts the focus towards the positive development and harmonization of software development teams. After exploring the complexities of leadership and management strategies in Part 2, this section delves into the practical aspects of nurturing a productive, collaborative, and growth-oriented environment within tech teams. The essence of this part lies in its emphasis on the development of both individuals and teams as a whole, recognizing that the success of tech projects is as much about technical skill as it is about effective teamwork and continuous personal and professional growth.

In the dynamic and often high-pressure world of software development, fostering an environment where team members can grow, collaborate effectively, and contribute to their fullest potential is crucial. This part addresses the strategies and practices that can enhance team dynamics, boost morale, and encourage a culture of continuous learning and improvement. It's about creating a workspace where innovation is nurtured, where diversity of thought and approach is valued, and where each team member feels empowered and motivated to contribute.

One of the key themes of this section is the recognition that growth and collaboration are interdependent. Personal growth in skills and competencies enables team members to contribute more effectively to team efforts, while a collaborative environment provides the support and stimulation necessary for individual development. This synergy between individual development and team collaboration is a central thread that runs through the chapters in this part.

Additionally, this section explores the importance of creating an inclusive team culture, one that embraces diverse perspectives and backgrounds. In the tech world, where teams are often composed of individuals from various cultures, experiences, and specializations, fostering an inclusive environment is not just a moral imperative but a practical necessity. It leads to richer ideas, more innovative solutions, and a more cohesive team dynamic.

Leadership's role in fostering this environment is also examined. Effective leaders in tech are those who can inspire their teams, provide clear direction and support, and create opportunities for growth and learning. They are adept at recognizing the unique strengths and potential of each team member and at orchestrating these diverse talents towards a common goal.

Cultivating a Growth-Oriented Mindset

Cultivating a growth-oriented mindset among developers is a vital strategy for fostering continuous improvement, adaptability, and innovation. In the fast-evolving field of software development, where new technologies and methodologies emerge regularly, a growth mindset – the belief that abilities and intelligence can be developed through dedication and hard work – is crucial. This mindset not only empowers developers to expand their skillsets and adapt to new challenges but also fosters a culture of personal responsibility and self-improvement within the team.

One of the primary strategies to foster a growth mindset is through the establishment of a learning culture within the team or organization. This involves creating an environment where continuous learning is valued, encouraged, and facilitated. Encouraging developers to dedicate time to learning new languages, tools, or methodologies can be achieved through various means, such as offering access to online courses, organizing in-house training sessions, or encouraging attendance at conferences and workshops. Recognizing and rewarding learning and improvement efforts can also reinforce the importance of ongoing personal and professional development.

Setting challenging but achievable goals is another effective way to promote a growth mindset. These goals should push developers slightly out of their comfort zones, encouraging them to stretch their abilities and learn new skills. Regular check-ins and feedback sessions can help ensure that these goals remain aligned with personal and organizational objectives, and provide opportunities for course correction and recognition of progress.

Mentorship and coaching play a significant role in developing a growth-oriented mindset. Pairing less experienced developers with more seasoned mentors can facilitate knowledge transfer, provide guidance, and offer support. This mentorship can be formal or informal but should be structured in a way that encourages open dialogue and mutual learning.

Promoting a culture of feedback is also crucial. Constructive feedback helps developers understand their strengths and areas for improvement. It's important that feedback is specific, actionable, and delivered in a manner that encourages development rather than defensiveness. Encouraging peer-to-peer feedback can also be beneficial, as it fosters a collaborative environment where team members learn from each other.

Encouraging reflection and self-assessment is another key strategy. Developers should be encouraged to reflect on their experiences, challenges faced, and lessons learned. This reflection can be facilitated through regular retrospective meetings or personal development plans. It helps developers take ownership of their growth and identify specific areas they want to develop.

Creating opportunities for developers to apply new skills and knowledge in practical projects is also vital. This application allows them to experiment, learn from real-world experiences, and see the impact of their growth on actual projects. It can be achieved by assigning them to new projects, different roles, or cross-functional teams.

Fostering Team Collaboration and Resilience

Fostering team collaboration and resilience in the face of entitlement issues is a critical challenge in the realm of software development. A collaborative, adaptable, and resilient team is not only more effective in achieving its goals but also better equipped to navigate the complexities and pressures of the tech industry. This chapter delves into strategies and practices that can build such teams, focusing particularly on overcoming the challenges posed by entitlement.

The foundation of a collaborative and resilient team is a strong sense of shared purpose and clear, common goals. Establishing a vision that resonates with all team members can unify the group, providing a clear direction and a sense of belonging. This shared purpose needs to be communicated effectively and reinforced regularly, ensuring that it remains at the forefront of the team's efforts.

Building trust among team members is another critical aspect of fostering collaboration and resilience. Trust is the glue that holds a team together, especially during challenging times. Creating an environment where team members feel safe to express their ideas, share their concerns, and be vulnerable is essential. This can be achieved through regular team-building activities, open communication channels, and an inclusive culture where every voice is valued.

Addressing entitlement directly is also crucial. Entitlement can manifest as a lack of willingness to cooperate, a resistance to feedback, or a sense of superiority over team members. Leaders must tackle these issues head-on by setting clear expectations about team behavior and performance. This involves establishing norms that promote equality, mutual respect, and a commitment to the team's objectives. It's important to recognize and celebrate collaborative efforts and to ensure that individual achievements do not overshadow the team's collective success.

Encouraging diversity of thought and fostering a culture of open-mindedness contribute significantly to team resilience and adaptability. Teams that embrace diverse perspectives are better equipped to solve problems creatively, innovate, and adapt to changing circumstances. This diversity goes beyond demographic differences and includes diversity in skills, experiences, and ways of thinking. Leaders should encourage team members to share their unique perspectives and create an environment where different viewpoints are not only accepted but actively sought out and valued.

Effective conflict resolution is another key element in building resilient teams. Conflicts, if not managed properly, can erode trust and cooperation. Leaders must develop the skills to mediate conflicts, ensuring that they are resolved in a manner that strengthens relationships and promotes understanding. This involves listening to all parties involved, understanding the underlying issues, and guiding the team towards a resolution that benefits the collective.

Providing opportunities for continuous learning and development also plays a vital role in fostering team resilience. When team members feel that they are growing and developing new skills, they are more engaged and committed to the team's success. Leaders should facilitate this growth by providing access to training, mentorship, and challenging projects that stretch the team's capabilities.

Developing a Consulting Mindset in Developers

Developing a consulting mindset in developers is pivotal in enhancing their ability to deliver value to clients. This mindset goes beyond technical proficiency, encompassing a client-focused approach that prioritizes problem-solving, understanding client needs, and ultimately ensuring client satisfaction. This chapter explores how to encourage developers to adopt this mindset, which is crucial for the success of projects, especially in environments where the end-user's satisfaction is paramount.

The first step in fostering a consulting mindset is to instill an understanding of the importance of the client's perspective. Developers need to appreciate that while technical excellence is critical, the primary goal is to solve the client's problems or meet their needs effectively. This understanding can be cultivated through regular interactions with clients, including participation in meetings and discussions where clients express their needs, concerns, and feedback. Such direct exposure allows developers to gain insights into the client's world, fostering empathy and a better understanding of how their work impacts the client.

Problem-solving is at the heart of a consulting mindset. Developers should be encouraged to think beyond code and consider the broader context in which their solutions will be applied. This involves understanding the business implications of their work, considering the user experience, and being proactive in identifying and addressing potential issues. Training in critical thinking and problem-solving techniques can be beneficial, as can workshops or sessions that focus on case studies and real-world scenarios.

Client satisfaction is another key aspect of a consulting mindset. Developers should be taught to view their work through the lens of client satisfaction, which often involves going the extra mile to ensure that the solutions provided not only meet the technical requirements but also align with the client's broader business objectives and user needs. This can be encouraged by setting project goals that are aligned with client satisfaction metrics and by recognizing and rewarding work that positively impacts the client.

Effective communication skills are also essential in developing a consulting mindset. Developers must be able to communicate their ideas and solutions clearly and effectively, both within the team and with clients. This involves not just verbal and written communication skills but also the ability to listen actively and interpret client needs accurately. Providing training and opportunities to develop these skills, such as presentations, client meetings, and feedback sessions, can be immensely beneficial.

Collaboration is another crucial component. A consulting mindset thrives in a collaborative environment where developers work closely with clients, project managers, and other stakeholders. Encouraging a culture of collaboration, where developers are part of cross-functional teams and are involved in various stages of the project, can help in developing a more holistic view of the project and the client's needs.

Balancing the Scales: Genius vs. Productive Developers

In the software development realm, team composition often oscillates between two distinct types of developers: the 'genius' developers known for their exceptional creativity and problem-solving skills, and the consistently productive developers who are reliable, steady, and efficient. Balancing these two types is crucial for creating a well-rounded team that can innovate while reliably meeting project deadlines and maintaining quality standards.

'Genius' developers often bring a level of innovation and creativity that can lead to groundbreaking solutions. Their ability to think outside the box, tackle complex problems, and come up with unique approaches is invaluable in a field driven by rapid technological advancement and intense competition. However, this brilliance can sometimes come with trade-offs. Genius developers may work in bursts of inspiration, which can lead to inconsistent productivity. Their focus on innovative solutions may sometimes overlook practical considerations or project constraints. Additionally, their strong convictions about their ideas can pose challenges for team cohesion and collaborative efforts.

On the other side of the spectrum are developers whose strength lies in their consistent productivity. These developers are the backbone of any team, known for their reliability, ability to meet deadlines, and maintain a steady output. They ensure that the day-to-day tasks of software development are completed efficiently and that the project progresses steadily. However, these developers might sometimes be less inclined to venture into uncharted territories or engage in high-risk, high-reward innovation.

Finding the right balance between these two types of developers is key to building a successful team. The following strategies can help in achieving this balance:

- Leveraging Strengths While Encouraging Growth: Recognize and utilize the unique strengths of each type of developer. Allow genius developers the space to innovate while channeling their creativity into productive outcomes. Simultaneously, encourage consistently productive developers to engage in creative problem-solving and explore new technologies, thus fostering a more dynamic skill set.

- Setting Clear Expectations and Goals: Establish clear project goals and milestones that cater to both innovation and productivity. This clarity helps genius developers align their creative efforts with project objectives while ensuring that consistently productive developers understand the broader vision and contribution of their steady work.

- Fostering a Collaborative Environment: Create a team culture that values diverse contributions and fosters collaboration. Encourage open communication and idea-sharing sessions where both types of developers can learn from each other, thus creating a more cohesive team dynamic.

- Flexible Project Management: Adopt a flexible approach to project management that accommodates different working styles. This might involve a mix of structured and unstructured tasks, allowing for both creative exploration and methodical progress.

- Regular Feedback and Adaptation: Implement a system of regular feedback that allows for continuous assessment and adaptation of team dynamics. Use this feedback to make adjustments in team composition, project planning, and management approaches to better balance the scales between genius and productivity.

- Recognizing and Rewarding All Contributions: Develop a recognition system that values both innovative solutions and consistent productivity. Acknowledging the different types of contributions reinforces their importance and motivates all team members.

Balancing the scales between genius talent and consistent productivity requires thoughtful team management, recognition of diverse strengths, and a flexible approach to project execution. By achieving this balance, teams can harness the benefits of both innovation and reliability, leading to successful, well-rounded software development outcomes.

Part 4: Navigating Challenges and Adapting to Change

Part 4 of this book, "Navigating Challenges and Adapting to Change," addresses the inevitable and often unpredictable obstacles that arise in the fast-paced world of software development. This section is particularly focused on the resilience and adaptability required to effectively manage and overcome these challenges. After exploring the aspects of team growth, collaboration, and individual mindset in the previous sections, this part shifts the focus towards external and internal hurdles that can impact a team's performance and the strategies to navigate through them.

In the ever-evolving landscape of technology, change is the only constant. Whether it's adapting to new technologies, navigating market shifts, managing team dynamics, or dealing with project setbacks, the ability to adapt and evolve is crucial for sustained success. This section delves into practical strategies for adapting to these changes, ensuring that teams remain agile, focused, and capable of turning challenges into opportunities for growth.

One key aspect of this part is understanding the nature of challenges in software development. These can range from technological advancements that render current skills obsolete, to shifts in project scope, to interpersonal conflicts within teams. Each of these challenges requires a distinct approach and mindset to address effectively. Leaders and team members alike must be equipped with the skills to identify and tackle these issues proactively.

Another critical theme in this section is the importance of resilience. In the face of setbacks and obstacles, the ability of a team to bounce back and continue moving forward is invaluable. This resilience is not just about enduring difficulties but also about learning from these experiences and using them to strengthen the team's capabilities and processes.

Moreover, this part discusses the significance of fostering an environment that embraces change rather than resists it. This involves creating a culture where experimentation, innovation, and risk-taking are encouraged, and failures are viewed as learning opportunities. Such an environment empowers teams to adapt to changes more readily and to approach challenges with a solution-oriented mindset.

Managing Entitlement in High-Pressure Environments

Managing entitled behavior in high-pressure environments, especially in the context of high-stakes software development projects, presents unique challenges. In such settings, the pressure to deliver exceptional results under tight deadlines can exacerbate entitlement tendencies among developers. These tendencies, if not managed effectively, can lead to conflicts, derail team dynamics, and ultimately compromise the project's success. This chapter explores strategies for addressing and mitigating entitled behavior in such demanding environments.

Recognizing the signs of entitlement is the first step in managing it. In high-pressure situations, entitled behaviors may manifest as resistance to feedback, reluctance to collaborate, insistence on special treatment, or an inability to adapt to changing project demands. These behaviors can stem from various sources, including stress, fear of failure, or a sense of superiority over team members. Identifying these behaviors early is crucial for timely intervention.

Clear communication of expectations is essential in managing entitlement. Leaders must articulate the project's objectives, the importance of teamwork, and the expectations for professional conduct under pressure. This clarity helps align team members' actions with the project's goals and sets a standard for acceptable behavior.

Promoting a culture of accountability is also vital. In high-pressure environments, it's important to ensure that all team members, regardless of their perceived status or contribution, are held to the same standards. This accountability can be reinforced through regular performance evaluations, feedback sessions, and, if necessary, disciplinary actions for unprofessional behavior.

Fostering an environment of support and understanding can help alleviate the stress that often fuels entitled behavior. Providing resources for stress management, encouraging open discussions about challenges, and showing empathy can create a more supportive and collaborative atmosphere. This approach can help team members feel valued and understood, reducing the likelihood of entitlement behaviors stemming from stress or insecurity.

Encouraging teamwork and collaboration is another key strategy. In high-pressure situations, fostering a sense of unity and shared purpose can help mitigate feelings of entitlement. Team-building activities, collaborative problem-solving sessions, and recognition of team achievements can reinforce the value of collective effort over individual accolades.

Leadership style plays a significant role in managing entitlement. Leaders in high-pressure environments should balance assertiveness with approachability. They need to be firm in addressing entitled behaviors while remaining open to listening to team members' concerns and perspectives. This balance helps maintain respect and authority while ensuring that team members feel heard and respected.

Finally, providing opportunities for personal and professional development can counteract entitlement. When team members are engaged in growth opportunities and challenged to improve their skills, they are more likely to focus on self-improvement rather than entitlement. This engagement can be facilitated through training programs, mentorship opportunities, and challenging assignments that push team members out of their comfort zones.

Managing entitlement in high-pressure environments requires a multifaceted approach that includes setting clear expectations, fostering a culture of accountability and support, encouraging collaboration, displaying balanced leadership, and providing development opportunities. By addressing entitlement effectively, leaders can maintain a productive, professional, and harmonious team environment, even under the most stressful and high-stakes circumstances.

The Hidden Cost: How Entitled Developers Can Undermine a Consultancy

Entitled behaviors in consultancy settings can pose significant risks, undermining the effectiveness and reputation of the consultancy. In a field where client satisfaction, adaptability, and teamwork are paramount, entitled attitudes can lead to a range of issues, from internal conflicts to diminished client trust. This chapter explores the risks associated with entitled behaviors in consultancy environments and strategies for mitigating these risks.

One of the primary risks of entitled behavior in a consultancy is the impact on team dynamics. Consultants are often required to work in diverse teams, sometimes collaborating with clients' in-house staff. Entitled behaviors, such as an unwillingness to cooperate, a disregard for team input, or an insistence on one's methods, can lead to conflicts, impede effective collaboration, and slow down project progress. This can not only affect the immediate project but also damage the long-term cohesion and effectiveness of the consultancy team.

Entitled behaviors can also negatively impact client relationships. Consultancies thrive on their reputation for professionalism and client-centric service. Entitled attitudes, such as a lack of flexibility, failure to acknowledge client concerns, or an inability to adapt to client feedback, can lead to dissatisfaction and erode trust. This can result in lost clients, negative reviews, and a damaged reputation, which can be particularly detrimental in the consultancy business where word-of-mouth and client testimonials hold significant weight.

To mitigate these risks, several strategies can be employed:

- Fostering a Client-Centric Culture: Cultivating a culture that prioritizes client needs and values adaptability is crucial. This involves regularly reinforcing the importance of client satisfaction and demonstrating, through leadership and policies, that entitled behaviors that compromise client service will not be tolerated.

- Setting Clear Expectations and Conduct Standards: Establishing clear guidelines on professional behavior, teamwork, and client interaction helps set the tone for expected conduct. Regular training sessions on professional ethics, client relations, and teamwork can reinforce these standards.

- Promoting Self-Awareness and Empathy: Encouraging consultants to develop self-awareness and empathy can reduce entitled behaviors. Workshops on emotional intelligence, role-playing scenarios, and feedback sessions can help consultants understand and appreciate the perspectives of others, fostering more cooperative and client-focused attitudes.

- Implementing Performance Feedback Mechanisms: Regular performance reviews that include feedback from peers, supervisors, and clients can help identify and address entitled behaviors. Constructive feedback, along with clear pathways for improvement, can guide consultants towards more professional and client-focused behaviors.

- Encouraging Collaboration and Team Building: Team-building exercises and collaborative projects can help break down silos and reduce entitled attitudes. These activities should be designed to promote mutual understanding, respect, and the value of diverse perspectives and skills.

- Providing Leadership Training: Equipping leaders and managers with the skills to recognize and address entitled behaviors is vital. Training in conflict resolution, communication, and leadership can empower them to effectively manage their teams and foster a more cooperative and client-focused environment.

- Offering Career Development Paths: Providing clear career advancement opportunities based on merit, professionalism, and client feedback can motivate consultants to adopt behaviors that align with the consultancy's values and goals.

While entitled behaviors can pose significant risks in consultancy settings, they can be effectively mitigated through a combination of culture shaping, clear expectations, training, feedback mechanisms, team-building activities, leadership development, and career advancement opportunities. By addressing these behaviors proactively, consultancies can maintain high standards of professionalism and client service, ensuring their success and reputation in the competitive consultancy landscape.

Melting the Snowflake Syndrome: Navigating Individualism

Navigating the challenges posed by highly individualistic developers, often characterized by the 'Snowflake Syndrome,' requires a nuanced approach that balances the need for individual expression with the imperative of team cohesion and collaboration. The key lies in managing these unique personalities in a way that leverages their strengths while integrating them effectively into the team. This chapter outlines strategies for handling such individualism in software development teams, ensuring that it contributes positively to team dynamics and project outcomes.

Recognizing and valuing individual strengths is the first step in managing highly individualistic developers. Each developer brings a unique set of skills, experiences, and perspectives to the table. Leaders should acknowledge and appreciate these individual contributions, understanding that they can add significant value to the team. This recognition can be achieved through regular one-on-one meetings, personalized feedback, and public acknowledgment of their contributions. By doing so, leaders can make these developers feel valued for their uniqueness, thereby reducing the potential for disruptive expressions of individualism.

Setting clear team goals and expectations is crucial. Highly individualistic developers need to understand how their work fits into the larger picture and contributes to the team's objectives. Leaders should clearly communicate the team's goals, the roles and responsibilities of each member, and the expectations for collaboration and teamwork. This clarity helps individualistic developers align their efforts with the team's objectives and understand the importance of working collaboratively.

Fostering a culture of collaboration is another vital strategy. This involves creating an environment where teamwork is valued and rewarded. Encouraging collaborative projects, pair programming, and cross-functional teams can help break down silos and promote a more cooperative work culture. Workshops and team-building activities focused on collaboration and communication skills can also be beneficial in cultivating a sense of unity and mutual respect among team members.

Encouraging open communication and feedback is essential in managing individualism. Creating a safe space for open dialogue allows individualistic developers to express their ideas and concerns, fostering a culture of transparency and inclusivity. Regular team meetings, feedback sessions, and open-door policies can facilitate this communication. Additionally, encouraging peer-to-peer feedback can help these developers understand how their behavior impacts others, promoting a more team-centric perspective.

Integrating individualistic developers into the team can also be achieved by assigning them roles that play to their strengths while requiring collaboration. For example, placing them in positions where their unique skills are needed for a team project can help them see the value of working with others. This integration should be done thoughtfully, ensuring that it aligns with their interests and skills.

Leaders should also lead by example. Demonstrating a balance between individual excellence and teamwork sets a powerful precedent. Leaders who show respect for individual contributions while emphasizing the importance of collaboration and collective success can inspire similar behavior in their teams.

Managing highly individualistic developers, or those exhibiting the 'Snowflake Syndrome,' requires a multifaceted approach that includes recognizing individual strengths, setting clear team goals, fostering a collaborative culture, encouraging open communication, thoughtfully integrating individuals into the team, and leading by example. By employing these strategies, leaders can harness the benefits of individualism while building cohesive, collaborative, and high-performing teams in the software development environment.

The Impact of Non-Development Tasks on Developer Morale

The balance between development and administrative tasks in the realm of software engineering is a critical factor that can significantly impact developer morale and, consequently, the overall success of projects. While development tasks are often viewed as the core and rewarding part of a developer's job, administrative duties such as documentation, time tracking, and attending meetings are equally essential for the smooth running and success of projects. This chapter examines how the balance between these two aspects of a developer's role affects their morale and the strategies to optimize this balance for project success.

Administrative tasks, though crucial, are frequently perceived by developers as tedious, time-consuming, and a diversion from the more engaging and creative aspects of coding. When developers are burdened with excessive administrative responsibilities, it can lead to frustration, a decrease in job satisfaction, and ultimately, a decline in morale. This reduction in morale can have ripple effects on productivity, quality of work, and team dynamics, potentially jeopardizing project timelines and outcomes.

On the other hand, a lack of adequate attention to administrative tasks can lead to disorganization, miscommunication, and a lack of accountability within projects. Essential elements such as thorough documentation, accurate time tracking, and effective meeting management are vital for maintaining project clarity, continuity, and alignment with client expectations. Neglecting these aspects can result in project delays, quality issues, and client dissatisfaction.

Balancing these two types of tasks requires a strategic approach. One effective method is the integration of administrative tasks into the development workflow in a manner that minimizes disruption and maximizes efficiency. This can be achieved through the use of project management tools that streamline administrative processes, automation of repetitive tasks, and setting aside specific times for administrative duties to reduce the interruption of the development flow.

It is also crucial to communicate the importance and relevance of administrative tasks to the overall success of projects. Understanding the bigger picture can help developers appreciate the value of these tasks and motivate them to engage in them more willingly. Highlighting how effective documentation, for instance, can ease future development efforts or how accurate time tracking can lead to better project planning and resource allocation can align developers' perception of these tasks with their critical role in project success.

Another strategy is to tailor the amount and type of administrative tasks to individual developers' preferences and strengths. Some developers may have a natural inclination towards meticulous documentation or enjoy the organizational aspect of project management. Leveraging these individual preferences can distribute administrative tasks more effectively, ensuring that they are handled efficiently while keeping developers engaged in activities that align with their interests and skills.

Providing training and resources to help developers manage administrative tasks more effectively can also improve morale. This includes offering training on time management, effective meeting strategies, and documentation best practices. Equipping developers with the skills to handle these tasks efficiently can reduce the perceived burden and improve their engagement with these aspects of their role.

The balance between development and administrative duties is a delicate one that significantly impacts developer morale and project success. By integrating administrative tasks into the development workflow, effectively communicating their importance, tailoring tasks to individual strengths, and providing the necessary training and resources, organizations can optimize this balance. This approach not only maintains high morale among developers but also ensures the smooth execution and success of software development projects.

Distance and Dynamics: Leading Remote and Hybrid Developer Teams

Leading remote and hybrid developer teams in the tech industry presents a unique set of challenges and opportunities. The shift to remote and hybrid work environments has fundamentally altered the dynamics of team interaction and collaboration. This chapter explores the specific challenges inherent in managing remote and hybrid teams and outlines effective strategies to address these challenges, ensuring that teams remain cohesive, productive, and motivated.

One of the primary challenges in leading remote and hybrid teams is maintaining effective communication. Without the benefit of face-to-face interactions, conveying complex technical ideas, and ensuring that all team members are on the same page can be difficult. Miscommunications can lead to errors, delays, and frustrations. To address this, it's essential to establish robust communication channels and protocols. Regular video conferences, instant messaging platforms, and collaborative project management tools can bridge the communication gap. Establishing regular check-ins and team meetings ensures ongoing alignment and provides opportunities for team members to connect and engage with each other.

Building and maintaining team cohesion is another significant challenge in remote and hybrid settings. The lack of physical presence and informal interactions can lead to feelings of isolation and disconnection among team members. Fostering a sense of team unity requires intentional effort. This can include virtual team-building activities, informal virtual coffee breaks, and creating online spaces for casual conversations and social interactions. Recognizing and celebrating team achievements, even in a virtual environment, can also strengthen the sense of community and belonging.

Managing productivity and ensuring accountability without micromanagement is crucial in remote and hybrid teams. Leaders need to trust their teams while establishing clear expectations and deliverables. Setting measurable goals, providing clear guidelines, and utilizing productivity tracking tools can help in managing team performance. However, it's important to balance this with trust and flexibility, recognizing that remote work can involve different work schedules and environments.

Another challenge is ensuring equity and inclusion in hybrid teams, where some team members are in-office while others are remote. There's a risk of remote employees feeling left out of decisions or informal communications that happen in the office. Leaders must be conscious of these dynamics and strive to provide equal access to information, opportunities, and interactions for all team members, regardless of their physical location.

Adapting leadership styles to suit remote and hybrid environments is also vital. Leaders must be more proactive in reaching out to team members, understanding their challenges, and providing support. Emotional intelligence becomes even more critical in remote settings, as leaders need to be attuned to non-verbal cues in virtual meetings and be sensitive to the nuances of written communications.

Providing support for personal development and career growth is equally important in remote and hybrid teams. This includes offering virtual training sessions, online professional development opportunities, and regular career development discussions. Ensuring that remote employees have the same opportunities for growth and advancement as their in-office counterparts is essential for maintaining motivation and job satisfaction.

Leading remote and hybrid developer teams requires a nuanced understanding of the challenges and a strategic approach to address them. By focusing on effective communication, team cohesion, productivity management, equity and inclusion, adaptable leadership, and support for personal and career development, leaders can successfully navigate the complexities of remote and hybrid work environments. These strategies ensure that teams remain connected, engaged, and productive, irrespective of their physical work locations.

Part 5: Looking Ahead: Anticipating and Adapting to Emerging Industry Trends

Part 5, titled "Looking Ahead: Anticipating and Adapting to Emerging Industry Trends," marks a forward-thinking conclusion to our exploration of managing software development teams. In this final section, we pivot our focus to the future, underscoring the importance of foresight and adaptability in the rapidly evolving tech industry. This part is dedicated to preparing leaders and teams not just to respond to changes but to anticipate and harness them proactively. It emphasizes the need to stay ahead of the curve in an industry where technological advancements and shifting market dynamics can render today's best practices obsolete tomorrow.

The tech industry is characterized by its relentless pace of change and innovation. New technologies, methodologies, and paradigms emerge continuously, each promising to be the next big disruptor. In such an environment, being reactive is no longer sufficient; teams and leaders must be anticipatory, recognizing emerging trends and preparing for their impact. This part delves into strategies for staying informed about industry developments, assessing their potential impact, and integrating them into team practices and project strategies.

A key theme of this section is the cultivation of a culture of continuous learning and adaptability within teams. In an industry where the only constant is change, the ability to learn and adapt is a critical competitive advantage. This part explores how to foster an environment that encourages experimentation, upskilling, and the embracement of new technologies and methods. It discusses the importance of nurturing a mindset that views change not as a threat but as an opportunity for growth and innovation.

Another crucial aspect is the strategic planning for future trends. This involves not only technical preparedness but also an organizational and cultural readiness to embrace change. Strategies for future-proofing teams, from diversifying skills to fostering a flexible and agile organizational structure, are discussed. This part also addresses the challenges of planning in the face of uncertainty, offering insights into scenario planning and risk management in the context of software development.

Emerging trends such as artificial intelligence, remote work dynamics, cybersecurity, and ethical considerations in software development are examined. This part provides a roadmap for understanding these trends, their potential impacts on software development practices, and the ways in which teams and leaders can adapt to these changes effectively.

Part 5 serves as a guide to navigating the future of software development. It equips leaders and teams with the knowledge, skills, and strategies required to anticipate and adapt to the ever-changing landscape of the tech industry. By embracing change, fostering a culture of continuous learning, and strategically planning for the future, teams can ensure that they not only survive but thrive in the face of the challenges and opportunities that lie ahead in the dynamic world of technology.

Anticipating and Adapting to Emerging Industry Trends

Anticipating and adapting to future shifts in developer attitudes, team structures, and tech industry trends is a critical task for leaders in the realm of software development. As the industry continues to evolve at a rapid pace, fueled by technological advancements and changing market dynamics, staying ahead requires a proactive and strategic approach. This chapter explores the methods and practices necessary for preparing and adapting to these imminent changes.

Understanding and preparing for shifts in developer attitudes is crucial in maintaining a motivated and effective workforce. As the tech industry evolves, so do the expectations and values of developers. Emerging trends such as a greater emphasis on work-life balance, remote working, and a focus on ethical and socially responsible coding are reshaping what developers expect from their employers and their work. To adapt to these changing attitudes, companies need to stay attuned to their workforce's evolving needs. This can be achieved through regular surveys, feedback sessions, and open forums. Adjusting policies to reflect these changing preferences, such as offering more flexible working arrangements or engaging in socially responsible projects, can help in attracting and retaining top talent.

The structure of development teams is also undergoing transformation. Traditional hierarchical structures are increasingly giving way to more agile, flat, and cross-functional teams. This shift calls for a change in management styles, from top-down leadership to more collaborative and facilitative approaches. Preparing for this change involves training leaders and managers to adopt these new styles, emphasizing skills like emotional intelligence, effective communication, and conflict resolution. Encouraging a culture of continuous learning and collaboration within teams is also essential, as it fosters adaptability and innovation.

Adapting to emerging tech industry trends requires staying informed and agile. Technologies such as artificial intelligence, machine learning, and blockchain are rapidly transforming the landscape of software development.

Keeping abreast of these trends through continuous learning and professional development is essential. This involves providing teams with access to training courses, workshops, and conferences. It also means encouraging a culture of experimentation and innovation where team members can explore new technologies and incorporate them into their work.

The integration of new technologies and methodologies into existing workflows can be challenging. It requires a careful assessment of the potential impact on current projects and processes. Pilot projects can be an effective way to test new technologies and methodologies before a full-scale roll-out. Regular review and adaptation of workflows and processes to incorporate new technologies are also crucial for staying competitive and efficient.

Ethical considerations and social responsibility are becoming increasingly important in software development. As technology becomes more ingrained in every aspect of society, the ethical implications of software development can no longer be overlooked. Preparing for this trend involves incorporating ethical considerations into the development process. This might include training on ethical coding, establishing ethics committees, or implementing review processes for ethical compliance.

Anticipating and adapting to future shifts in developer attitudes, team structures, and tech industry trends requires a multifaceted approach. It involves staying attuned to the evolving needs and expectations of developers, adopting more flexible and collaborative team structures, keeping abreast of emerging technologies, and integrating ethical considerations into the development process. By embracing these changes and preparing for them proactively, leaders and teams in the software development industry can ensure they remain effective, relevant, and competitive in the rapidly evolving tech landscape.

Generational and Cultural Diversity

Embracing generational and cultural diversity within tech teams is not only a moral imperative but also a strategic advantage. In an industry driven by innovation and problem-solving, the varied perspectives and experiences that come with a diverse team can be a significant asset. This chapter explores the importance of generational and cultural diversity in tech teams and how leveraging these diverse strengths can lead to greater success and innovation.

Generational diversity in tech teams brings together a blend of experiences and perspectives that can enhance creativity and problem-solving. Older team members often bring a wealth of experience, industry knowledge, and a long-term perspective on technology's evolution. They can offer insights into past tech trends and provide a measured approach to project management and decision-making. Younger team members, on the other hand, often bring fresh ideas, proficiency with new technologies, and a pulse on the latest industry trends. They are typically more adept at adapting to new tools and methodologies and often bring an energy and eagerness to innovate.

Cultural diversity adds another layer of richness to a team. Team members from different cultural backgrounds can introduce varied approaches to problem-solving, communication, and collaboration. They bring unique viewpoints shaped by their cultural experiences, which can lead to more innovative and well-rounded solutions. In an increasingly globalized market, having a team that understands and reflects this diversity can be a significant advantage, especially when developing products and services for diverse user bases.

To leverage the strengths of generational and cultural diversity, it's crucial to foster an inclusive environment where all team members feel valued and respected. This involves acknowledging and celebrating the different backgrounds and experiences that each member brings to the team. Inclusive practices can include regular diversity and inclusion training, creating forums for sharing cultural experiences, and implementing policies that promote equality and respect.

Effective communication is key in diverse teams. Cultural and generational differences can sometimes lead to misunderstandings or misinterpretations. Clear, open communication, along with an effort to understand and respect these differences, can mitigate potential conflicts. This might involve adapting communication styles to suit different team members or providing training on effective cross-cultural communication.

Another strategy to harness the power of diversity is through diverse team composition. When forming teams or assigning roles, it's beneficial to consider a balance of generational and cultural backgrounds. This deliberate composition ensures that a range of perspectives is represented and that team members can learn from each other's experiences and viewpoints.

Encouraging mentorship and knowledge sharing within the team can also be a powerful way to leverage generational and cultural diversity. Pairing younger team members with more experienced colleagues for mentorship can facilitate a two-way exchange of knowledge and skills. Similarly, cultural exchange programs or initiatives can enhance understanding and appreciation of different cultural backgrounds.

In conclusion, generational and cultural diversity in tech teams offers a wealth of benefits, from enhanced creativity and problem-solving to a better understanding of global markets. By fostering an inclusive environment, promoting effective communication, strategically composing teams, and encouraging mentorship and knowledge sharing, leaders in the tech industry can leverage these diverse strengths for greater team success and innovation. Embracing this diversity is not just about creating a fair and respectful workplace; it's about tapping into a rich source of ideas, perspectives, and experiences that can drive the tech industry forward.

Building a Sustainable Tech Team Culture

Creating a sustainable, inclusive, and productive team culture in the tech industry is a multifaceted endeavor that goes beyond mere technical expertise. It involves cultivating an environment where all team members feel valued, respected, and empowered to contribute their best work. This chapter focuses on strategies for building such a culture, ensuring long-term success and a positive work environment.

The foundation of a sustainable team culture is inclusivity. An inclusive culture is one where diversity is not just acknowledged but celebrated. It's a culture where differences in background, perspective, and experience are seen as assets rather than obstacles. To foster this, leaders must actively work to create an environment of openness and respect. This can be achieved through diversity and inclusion training, open forums for discussion, and policies that promote equal opportunities for all team members. Inclusivity also means ensuring that all voices are heard and valued, which can lead to more innovative solutions and a stronger team.

Communication is a critical component of a productive team culture. Establishing clear, open lines of communication ensures that team members are informed, engaged, and aligned with the team's goals and objectives. Regular team meetings, one-on-one check-ins, and open-door policies help facilitate this. Additionally, providing channels for feedback and ensuring that it is acted upon can strengthen trust and show team members that their opinions matter.

Empowerment and autonomy are also key to a sustainable culture. When team members feel empowered to make decisions and take ownership of their work, they are more engaged and motivated. This can be encouraged by providing clear expectations, the necessary resources, and then trusting team members to execute. Empowerment also means providing opportunities for professional growth and development, such as training, mentorship programs, and career advancement paths.

Promoting work-life balance is essential in the tech industry, known for its fast pace and high-pressure environments. Encouraging reasonable work hours, flexible working arrangements, and providing support for personal commitments can help prevent burnout and maintain high morale. A culture that values work-life balance not only improves job satisfaction but also attracts and retains top talent.

Team building and social activities play a significant role in fostering a cohesive and friendly work environment. Regular team outings, social events, and group activities can help build rapport, improve communication, and create a sense of camaraderie among team members. These activities should be inclusive and consider the diverse interests and backgrounds of the team.

Finally, a culture of continuous learning and adaptability is crucial in the ever-evolving tech landscape. Encouraging and providing resources for continuous learning, staying abreast of industry trends, and being open to change are essential. This involves not just technical learning but also developing soft skills like leadership, communication, and problem-solving.

Building a sustainable, inclusive, and productive team culture in the tech industry requires a holistic approach. It involves creating an inclusive environment, fostering open communication, empowering team members, promoting work-life balance, engaging in team-building activities, and nurturing a culture of continuous learning and adaptability. By implementing these strategies, leaders can create a team culture that not only drives success and innovation but also makes the workplace a positive and fulfilling environment for all team members.

Leading with Insight and Adaptability

In the dynamic and ever-evolving world of technology, the keys to effective project management and leadership lie in adaptability, continuous learning, and insightful decision-making. As we conclude this exploration, it's crucial to reflect on the key learnings that have emerged and how they can be applied to navigate the complex landscape of tech project management.

Adaptability has emerged as a non-negotiable trait for success in the tech industry. The rapid pace of technological change and the evolving nature of market demands require teams and their leaders to be flexible and responsive. This adaptability isn't just about staying abreast of the latest technologies or methodologies; it's about cultivating an ability to pivot strategies, re-evaluate goals, and modify approaches in response to new information and changing circumstances. It involves being open to experimentation, learning from failures, and being willing to embrace change.

Continuous learning is another critical aspect that underpins successful tech project management. In an industry defined by innovation, the only way to maintain a competitive edge is through ongoing education and skill development. This commitment to learning should not be confined to technical knowledge alone but should also encompass soft skills like communication, leadership, and emotional intelligence. Continuous learning should be an integral part of a team's culture, with leaders setting the tone by actively engaging in their own professional development and encouraging their team members to do the same.

Insightful leadership is the cornerstone of effective tech project management. Insightful leaders understand not only the technical aspects of their projects but also the human elements. They are skilled at reading team dynamics, understanding individual motivations, and navigating complex client relationships. Such leaders make decisions that are not only data-driven but also empathetic and considerate of the broader impact on people and the project. They balance the need to achieve technical excellence with the necessity of maintaining a positive, productive team environment.

The importance of building a sustainable, inclusive team culture cannot be overstated. In an industry where project teams are often diverse and geographically dispersed, creating an environment where every member feels valued and included is crucial. This involves recognizing and leveraging the strengths of generational and cultural diversity, fostering open and respectful communication, and ensuring that all team members have equal opportunities to contribute and grow.

Leading with insight and adaptability in tech project management is about much more than delivering projects on time and within budget. It's about creating a team culture that values adaptability, fosters continuous learning, and is led by insightful, empathetic leaders. These elements are essential for navigating the challenges and capitalizing on the opportunities presented by the ever-changing tech landscape. By embracing these principles, tech leaders and project managers can drive innovation, achieve project success, and build teams that are resilient, engaged, and equipped to face the future.

Conclusion

As we close the final pages of Managing Developers in the Age of Entitlement: A Tech Leader's Guide we reflect on the essential insights and strategies explored throughout this book. The journey through the chapters has been an enlightening exploration of the challenges and opportunities inherent in managing and leading tech teams in today's fast-paced and ever-evolving industry. This book has aimed to equip project managers and leaders with the tools and understanding necessary to navigate the complex and dynamic landscape of technology development.

From the outset, we delved into the nuances of developer entitlement and the impact of industry giants on shaping team dynamics. We then navigated through the challenges of managing remote and hybrid teams, adapting leadership styles, and fostering growth and collaboration in diverse environments. Each chapter provided practical strategies for dealing with these varied aspects, emphasizing the importance of adaptability, continuous learning, and insightful leadership.

A recurring theme has been the significance of building a sustainable, inclusive, and productive team culture. We have seen how embracing generational and cultural diversity enriches teams, fostering an environment where different perspectives lead to innovation and creative problem-solving. The book has underscored the importance of understanding and adapting to emerging trends in software development, preparing teams for future shifts in technology and market demands.

Managing Developers in the Age of Entitlement: A Tech Leader's Guide serves as a comprehensive guide for those who lead and manage developer teams in the tech industry. The lessons and strategies presented are designed to guide project managers and leaders through the complexities of modern tech project management. The journey in tech leadership is continuous and ever-changing, and this book is a companion for those who aspire to lead with insight, adaptability, and a deep understanding of the diverse nature of their teams. By embracing the principles and practices discussed, leaders and managers can steer their teams toward success, fostering environments where innovation thrives and where every team member feels valued and empowered to contribute their best work.